CW01395164

Looking at Genetics

Nina Sully

Batsford Academic and Educational London

LONDON BOROUGH OF BRENT
SCHOOL LIBRARY SERVICE

© Nina Sully 1985
First published 1985

All rights reserved. No part of this publication
may be reproduced, in any form or by any means,
without permission from the Publisher

Typeset by Tek-Art Ltd, West Wickham, Kent
and printed in Great Britain by
R.J. Acford
Chichester, Sussex
for the publishers
Batsford Academic and Educational,
an imprint of B.T. Batsford Ltd,
4 Fitzhardinge Street
London W1H 0AH

ISBN 0 7134 4775 3

Contents

Acknowledgment

The Author and Publishers thank the following for their kind permission to reproduce copyright illustrations: Heather Angel, pages 13, 19, 40; Barnaby's Picture Library, page 31; John Campbell, pages 16, 20, 24, 26, 29, 43; Richard and Sally Greenhill, page 7; Philip Harris Biological Ltd, page 22; Marc Henrie, page 11; Mencap, page 41; National Institute for Medical Research, page 14; Royal Agricultural Society of England, page 37; Royal Postgraduate Medical School, Hammersmith Hospital, London, page 35; R.S.P.B./E.A. Janes, page 12; Harry Smith, pages 18, 38/9; Syndication International, page 9; United Nations/Y. Nagata, page 33; *Wembley Observer*, page 45; Dr S. Willasden, A.F.R.C. Institute of Animal Physiology, Cambridge, page 44; Sarah Wyld/ Photo Co-op, page 8. The pictures were researched by Kate Fraser; the diagrams were drawn by the author.

Introduction

Everybody is special – a unique combination of characteristics that sets him or her apart from everyone else. This is perhaps surprising, since all humans have the same basic features: two eyes, a nose, a mouth, two arms and legs, and so on. The things which make someone a human being, and those which make him different from all other human beings, are all the result of the same process. They are passed on, or inherited, from the parents.

Genetics is the study of this process of inheritance. It tries to unravel the complicated methods by which characteristics are passed on from parents to their children. These methods are not unique to humans. Most animals and plants show remarkably similar patterns of inheritance.

Geneticists are scientists whose job is to find out more about inheritance. To do this they have to do experiments, breeding from particular, chosen animals and plants and seeing what sort of 'children', or offspring, they produce. Geneticists also have to study animals and plants under the microscope, to see exactly which structures are involved in inheritance and how they work. Another way of studying genetics is to look at populations of animals and plants and work out how they inherited their characteristics.

But genetics is not only used by geneticists. Plant and animal breeders use it to develop the best breeds. They are continually trying to breed better crops or farm animals, or to produce more showy flowers or dogs with perfect features. Genetics can be used in the fight against diseases. For instance, if we could understand how some viruses and bacteria inherit their disease-causing ability, we might be able to change them in some way and make them harmless. Many diseases, such as muscular dystrophy and cystic fibrosis, are inherited. The more we know about these diseases, through the study of genetics, the more likely we are to find a cure.

Family Connections

A family photo such as this often sparks off such comments such as "Little Jimmy has his grandad's nose" or "Mary is the spitting image of her mum". In fact, a quick glance at the faces in this family group reveals that, although each person is unique in appearance, certain features have been handed down from grandparents and parents to their children.

Baby humans, like many other animals, are the result of sexual reproduction. This means that two individuals, a male and a female, have produced another individual, their offspring. In this way, the species does not die out and continues from generation to generation.

During sexual reproduction, a complex series of events takes place during which material is passed from the parents to the offspring. This material forms a set of "instructions" which tell the offspring how to grow. On the one hand, the instructions tell the offspring to grow up to look like its parents, but on the other, they allow for variations which make the offspring unique. So, although the baby in the photograph has all the features of every normal human being, he is also an individual, distinguishable from every other human being.

The passing on of features, or characteristics, from parents to their offspring is called heredity. Children inherit a huge number of characteristics, from obvious things such as eye and hair colour, to more obscure things, such as personality and intelligence. There are also some inherited diseases.

The scientific study of heredity began in the middle of the last century when an Austrian monk, Gregor Mendel, experimented with varieties of peas grown in the kitchen gardens of the monastery. He worked out a set of laws of heredity which form the basis of a whole branch of modern biology known as genetics. These days, genetics involves working with lots of sophisticated equipment, as well as doing experiments with animals and plants in much the same way as Mendel did.

Genetics is not only fascinating, it is useful as well. Ever since man began domesticating animals and growing crops, he has been selecting the best individuals from which to breed. Gradually, we have built up a stock of animals and plants which provide us with food and other materials, which work for us, and which give us pleasure. In the past, animal and plant breeding was a rather hit or miss affair. It was often more by luck than judgement that a better animal or plant was produced. Nowadays, our knowledge of heredity can cut out some of the chance, and help us breed better domestic animals and plants.

Reproduction

This baby is ensuring the survival of the human race. In order to survive, all animals and plants must be able to reproduce, to make more of their own kind. Some plants and animals reproduce just by splitting in half or by growing an extra part of themselves, which then separates as a new individual. You may have a spider plant at home and noticed the new "baby" plants growing on the end of long stems. The "baby" plants will eventually grow roots and separate from their parent. This is asexual reproduction, where there is only one parent.

Perhaps the simplest sort of asexual reproduction is when the plant or animal just splits in two. Some of the tiniest animals are made of only one cell. *Amoeba* is a single-celled animal which lives in the muddy bottom of stagnant ponds. It has a central dark area called the nucleus, surrounded by a jelly-like grey mass called cytoplasm. Normally it is a constantly changing irregular shape but sometimes it stops moving and becomes spherical. Then the nucleus divides into two, the halves separate and the cytoplasm pinches into two as well. The *Amoeba* has reproduced, making two separate cells, just like the first.

A slightly more complicated sort of asexual reproduction is called "budding". A tiny animal, called *Hydra*, which lives in freshwater ponds, reproduces by budding. *Hydra* looks a bit like one of its relatives, the sea anemone, with a short stalk with tentacles at the top. Sometimes, a bump appears on the stalk, which grows bigger and develops tentacles at the free end. Finally, this "bud" breaks away, forming a new, separate *Hydra*.

If you grew the new *Hydra* in exactly the same conditions as its parents you would eventually be unable to tell them apart. The offspring of asexual reproduction are always exactly like their parents. The only way they might differ would be if their growth was affected by surrounding conditions, such as temperature, lack of food or attack by an enemy. The reason for this similarity between parent and offspring lies in the instructions handed on from the parent to its young. Since there is only one parent, only one set of instructions can be passed on to the offspring and these instructions can, therefore, never vary.

The babies in the maternity ward are the result of sexual reproduction. Each baby has two parents who both contributed to its

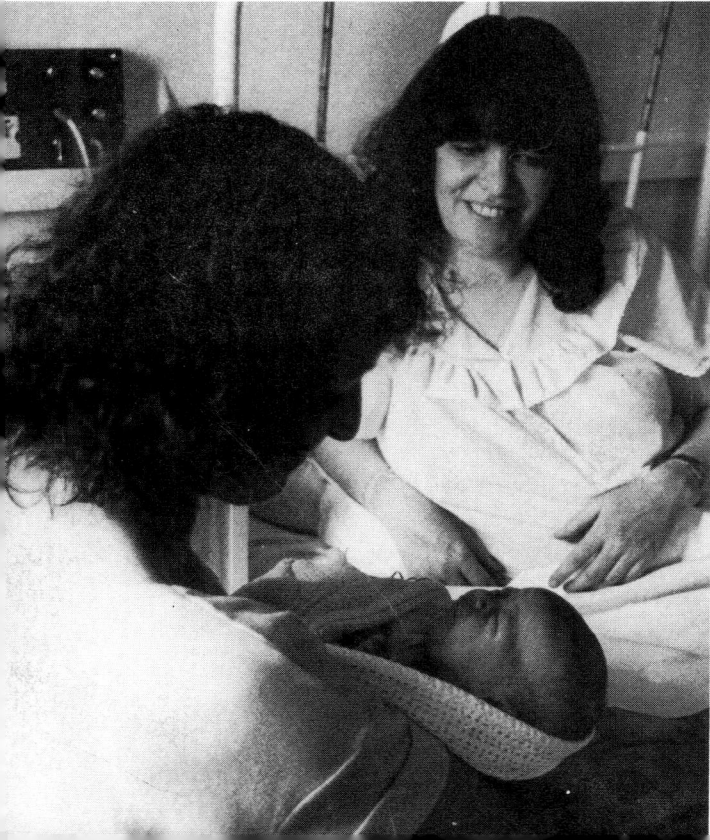

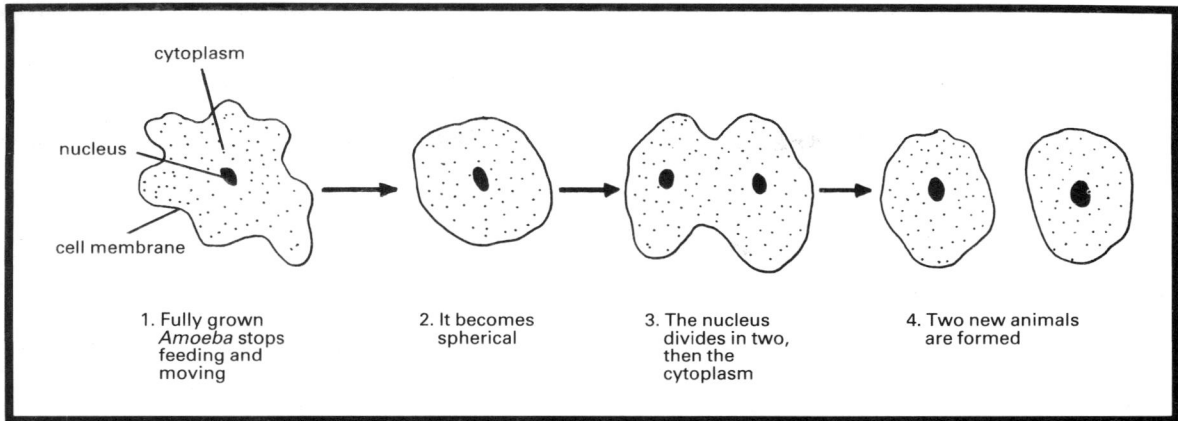

Asexual reproduction of amoeba

production. In most cases of sexual reproduction there is a male parent and a female parent, although in some very simple animals and plants, the differences between the two parents are so small that they are just called "plus" and "minus".

The offspring of sexual reproduction are all unique individuals. Each parent passes on a little of itself to its offspring which is therefore a mixture of its mother and its father. The visitors to the maternity ward have probably noticed that the baby has "his father's eyes" or "her mother's nose". This exact combination of father's and mother's features couldn't happen again, so the baby is unlike any other human being, even his brothers and sisters. These sextuplets are all unique, even though they are the same age and sex. Although they look similar and have features in common with both parents, they are all little individuals with their own identities.

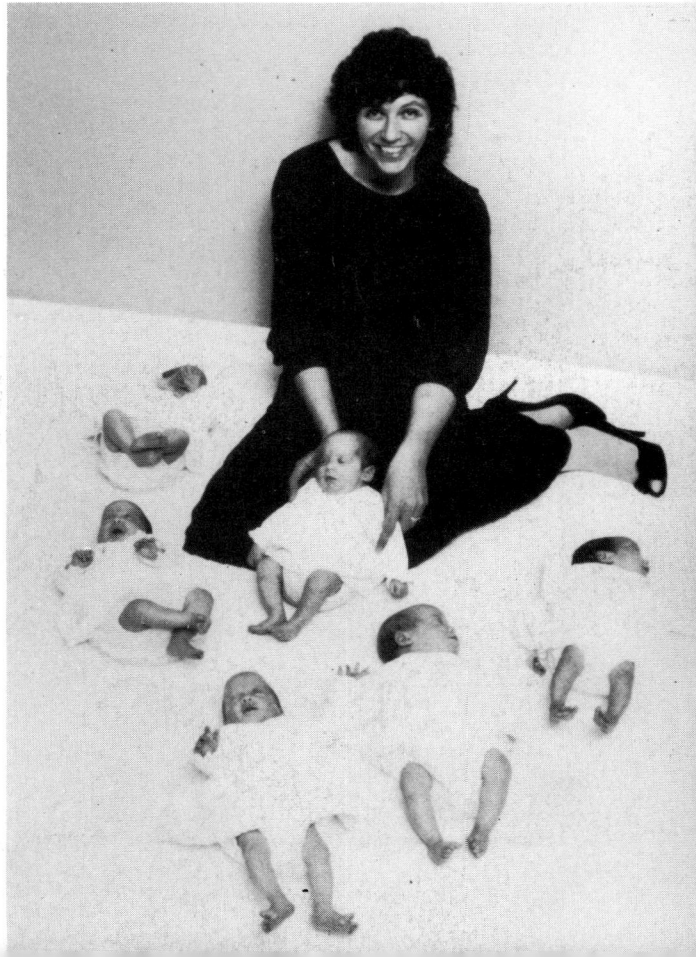

Sexual Reproduction

Not long ago these kittens were no bigger than a full stop. Twelve weeks before their birth their mother mated and began a complicated series of events known as sexual reproduction. Surprisingly, sexual reproduction follows a very similar pattern throughout the animal and plant kingdoms.

It is called sexual reproduction because it involves the two sexes, male and female. There are special sex organs inside the male and female, which make cells that begin the process of reproduction. These sex cells are called gametes. When a new life begins, one female gamete joins with one male gamete during a process called fertilization. The resulting cell, or zygote, divides into two, four, eight, 16 and so on, slowly developing into a new individual.

In most animals, the male sex organs are called testes and the male gametes called sperm. The female gametes in animals are called ova and they are made in the sex organs, or ovaries. In some animals, such as fish and frogs, the sperm and ova are shed from the male and female and join together outside their bodies. This is external fertilization. In other animals, such as birds and reptiles, sperm are passed into the female's body to join with the ova there. This is called internal fertilization. After fertilization, the zygotes of birds and reptiles develop into eggs which are laid outside the female's body. Inside the egg a young animal, called an embryo, grows and develops until it can survive outside the shell, when it hatches as a nestling bird or a young reptile. In mammals, such as the cat, sexual reproduction has gone one step further. Here, the embryo develops in the safety of the mother's body, fed and protected inside a special organ, the womb.

Of all animals, mammals have the most highly developed form of sexual reproduction. In the plant kingdom, it is the flowering plants

The steps in sexual reproduction

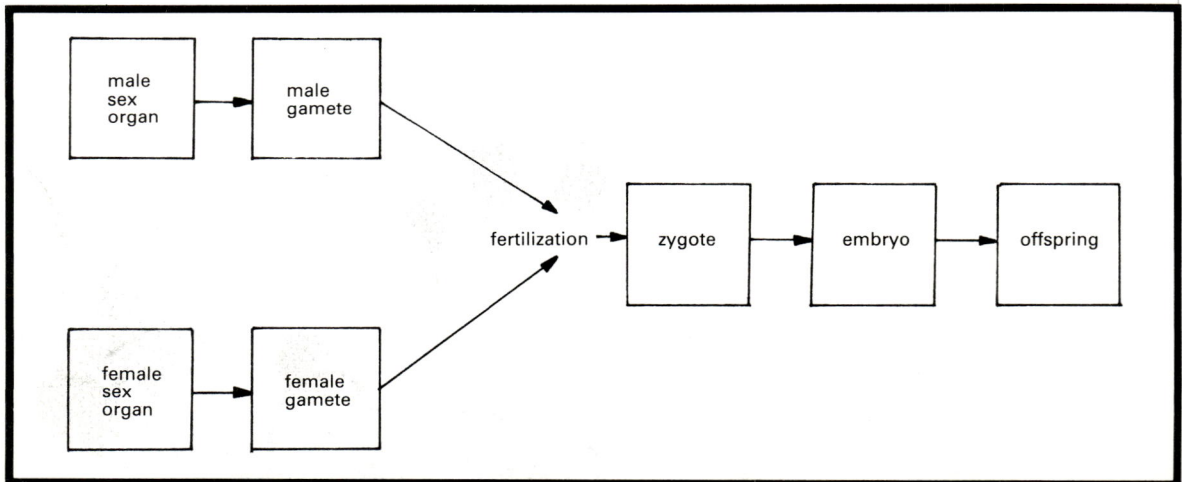

which are the most complex. The sex organs are gathered together in the flowers. Right in the middle of the flower are the female sex organs, or ovaries. Inside each ovary is a female gamete, or ovule. The male sex organs are the anthers, which produce thousands of male gametes, or pollen grains. The anthers split open, releasing pollen grains which are carried, by wind or insects, to the ovaries: a process called pollination. Fertilization, or the joining of a pollen grain with an ovule, happens inside the ovary. The resulting zygote develops into an embryo inside a seed which protects and nourishes it. The seeds are scattered away from the parent to grow into more plants if the conditions are right.

In spite of all these variations, sexual reproduction shows the same stages: male and female gametes are formed in the sex organs, they join at fertilization and develop into an embryo, which eventually grows into a new animal or plant. But how does the offspring inherit characteristics from its parents? To discover this, we need to look at the only things which pass from parents to offspring – the gametes.

Sex Cells

These blackbird's eggs started life as sperm and ova inside their parents' bodies. The sperm is the male gamete which is similar to the sperm of many other animals, including man. It consists of a head and a tail. Inside the head is a dark blob called the nucleus, which is vital to the process of reproduction. The tail is very mobile; it lashes like a whip, so that the sperm can swim in water. After mating, it swims to join the ovum in the female's body at fertilization. The sperm is very tiny; the head is only five thousandths of a millimetre across.

The ovum, or female gamete, is much bigger. It also has a nucleus, but it doesn't have a head and tail. Instead, it contains all it needs to develop into a baby bird. The yolk contains food to nourish the embryo, the white provides water and mineral salts and the shell protects the developing bird. In mammals, the embryo develops inside the mother's body, and the jelly-like material, or cytoplasm, surrounding the

The gametes of a mammal

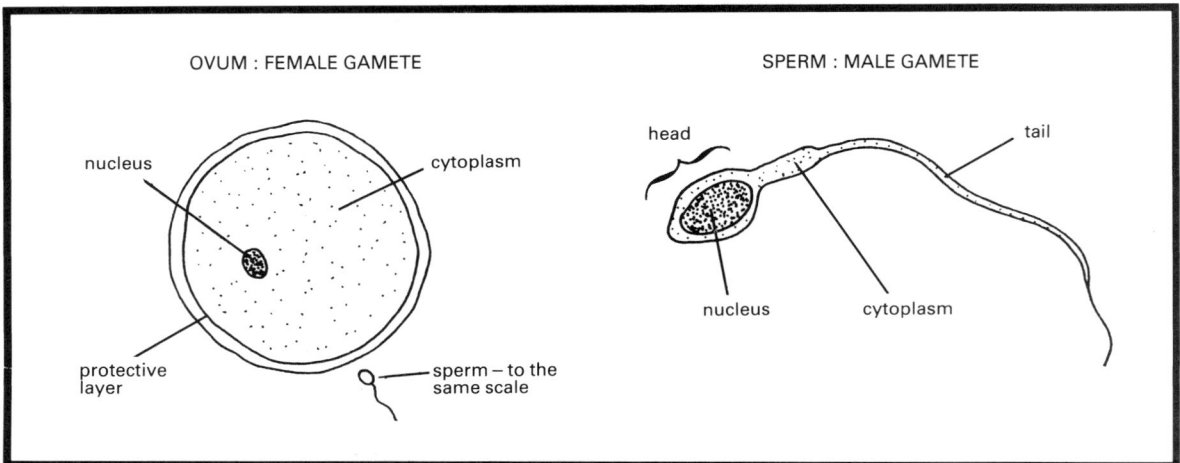

OVUM : FEMALE GAMETE

nucleus

cytoplasm

protective layer

sperm – to the same scale

SPERM : MALE GAMETE

head

tail

nucleus

cytoplasm

nucleus, contains just enough food to nourish the embryo until the mother takes over this job.

In flowering plants, such as these silver birch catkins, the male gametes, or pollen grains, form a yellow dust which is shed from the anthers of the flowers. Under the microscope, a pollen grain seems to have a simple structure. Like the animal sperm, it contains a dark blob, the nucleus. This is surrounded by a semi-transparent material, the cytoplasm, and the whole cell is protected by a thick, resilient cell wall. The pollen is carried to another flower by the wind or by insects and sticks to a special organ, the stigma, at the top of the ovary. Each pollen grain then develops a pollen tube which grows down into the ovary towards the ovule. The nucleus travels down the tube to join with that in the ovule.

The female gamete, or ovule, has a complicated structure, but, like all sex cells, it contains a nucleus which joins with the pollen grain's nucleus. It is surrounded by nourishing and protective materials, rather like the bird's egg. After fertilization a seed develops containing an embryo capable of growing into a new plant if given the right conditions.

All these gametes have one thing in common – the nucleus. Cells without a nucleus do not reproduce, so it is obvious that this tiny speck of material is vital for reproduction. What is really amazing about the nucleus is that, tiny as it is, it contains a vast store of information necessary for making a new animal or plant. Obviously, we need to take a closer look at the nucleus if we are to understand how this information is carried from parents to offspring.

The gametes of a flowering plant

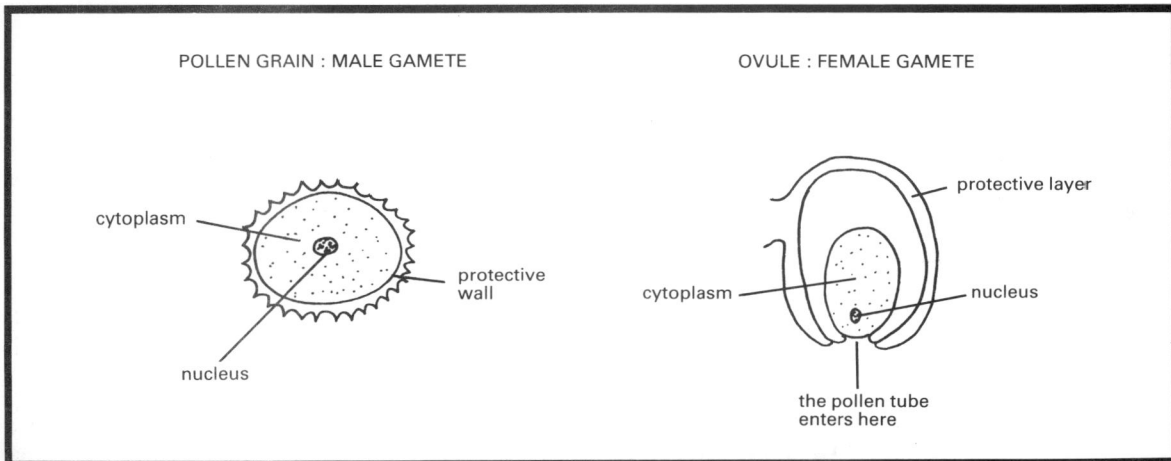

POLLEN GRAIN : MALE GAMETE

cytoplasm

protective wall

nucleus

OVULE : FEMALE GAMETE

protective layer

cytoplasm

nucleus

the pollen tube enters here

Inside the Nucleus

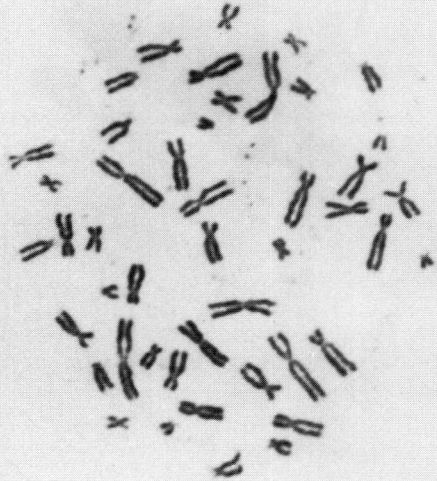

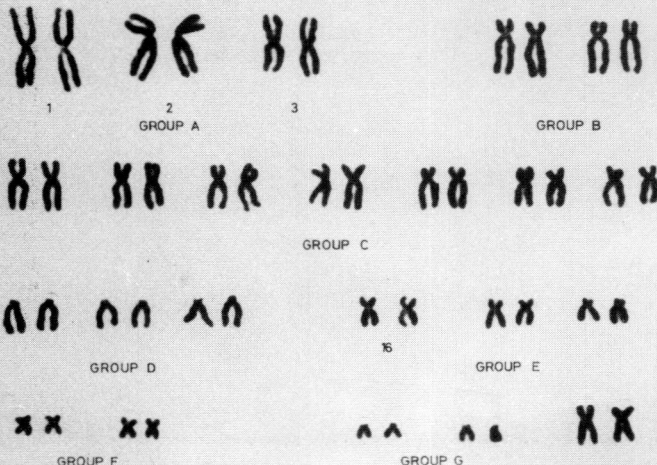

NORMAL FEMALE KARYOTYPE

1 2 3
GROUP A

GROUP B

GROUP C

GROUP D

16

GROUP E

GROUP F

GROUP G

An ordinary human nucleus looks like this under a powerful microscope. What appeared as a dark blob at lower magnifications is found to contain these separate threads, called chromosomes. If you count the chromosomes, you will find there are 46. No matter how many human nuclei you look at, so long as the person is normal, you will always count 46 chromosomes, except in the gametes.

You can often identify an animal or plant, just by looking at its nucleus. This is because every species has a set number of chromosomes in each nucleus. Normal rat cells, for instance, contain 42 chromosomes, fox cells have 34 chromosomes, and tomato cells have 24 chromosomes.

Not only do all animals or plants of the same species contain the same number of chromosomes in their cells, they have the same shapes and sizes of chromosomes as well. If we take a photograph of a human nucleus, and cut out the separate chromosomes, we can arrange them in a way which makes them easy to study. These chromosomes from a female human's cell have been arranged like this. We can count six of the longest chromosomes, four slightly shorter, 16 medium-sized ones, 12 small ones and eight very small ones. Notice also, that the chromosomes are arranged in pairs, with each member of pair looking exactly like the other. So, in fact, human cells have 23 *pairs* of chromosomes, rats 21 *pairs*, foxes 17 *pairs* and tomatoes 12 *pairs*.

Chromosomes float freely inside the nucleus, but at certain times they move in a very precise way. An animal or plant grows when its cells divide in two. The new cells increase in size and then divide in two again. This process continues and the organism grows. Just before the cell divides, each chromosome splits along its length

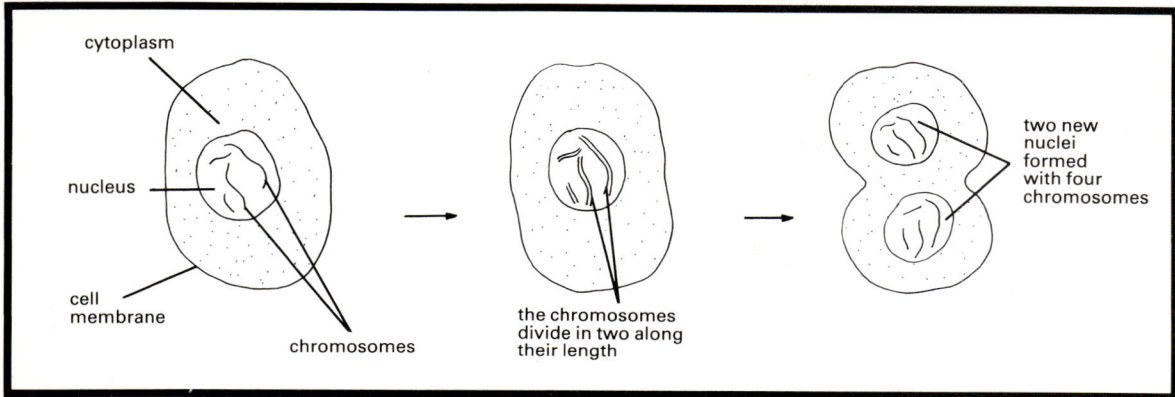

How cells divide when an animal or plant grows, in a cell with four chromosomes

to make two new chromosomes exactly like itself. You can see this happening in the photographs of human chromosomes. One of these chromosomes moves into one of the new cells and the other chromosome moves into the other. The result is two new cells, each with exactly the same number and types of chromosomes as the original cell. Obviously, it is important to the organism, that each of its cells contains exactly the right chromosomes.

The sex cells, or gametes, are rather different from all the other cells. If the gametes contained the same number of chromosomes as the normal cells, a problem would arise at fertilization. For instance, if a human sperm with 46 chromosomes joined with a human ovum also with 46 chromosomes, the zygote would contain 92 chromosomes and would no longer be a human cell. In fact, when sex cells are made, the number of chromosomes is halved; to 23 in human gametes. Then, when the two gametes join, the normal number is again restored. This means that the zygote receives half its chromosomes from its mother, and half from its father. We shall see that it is this mixture of chromosomes which makes us all unique individuals.

How cells divide to form gametes, in a cell with four chromosomes

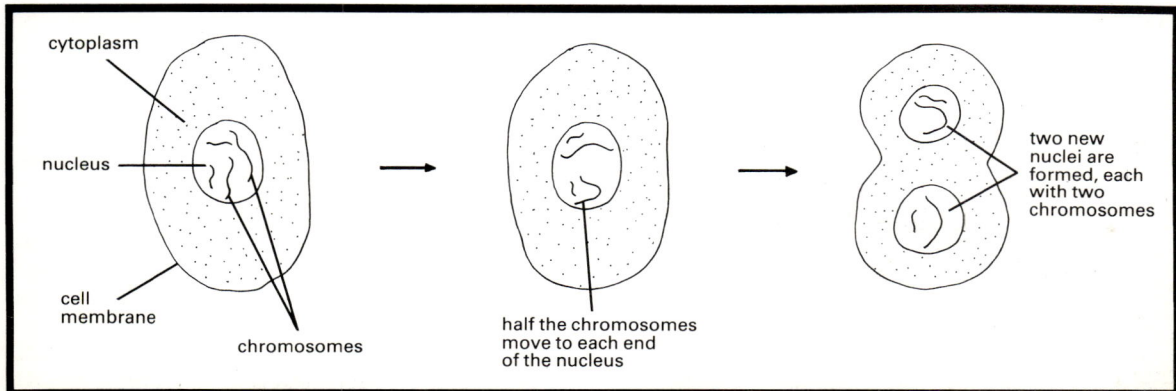

Genes

If you didn't know, you might think that these animals belonged to totally different species. There is such a variety of size, shape and colour, that it seems amazing that they are all dogs. To find out how this enormous variation has come about, we must look more closely at the chromosomes.

Exactly how chromosomes can produce such variety is not yet fully understood. Geneticists have found out some things about them, and can guess at the rest. They believe that each chromosome is divided along its length into units called genes. Each gene controls one characteristic of the individual, such as the colour of a dog's coat, or the length of its ears. Genes controlling the same characteristic are positioned in the same place on each member of a pair of chromosomes. So, if you laid a pair of

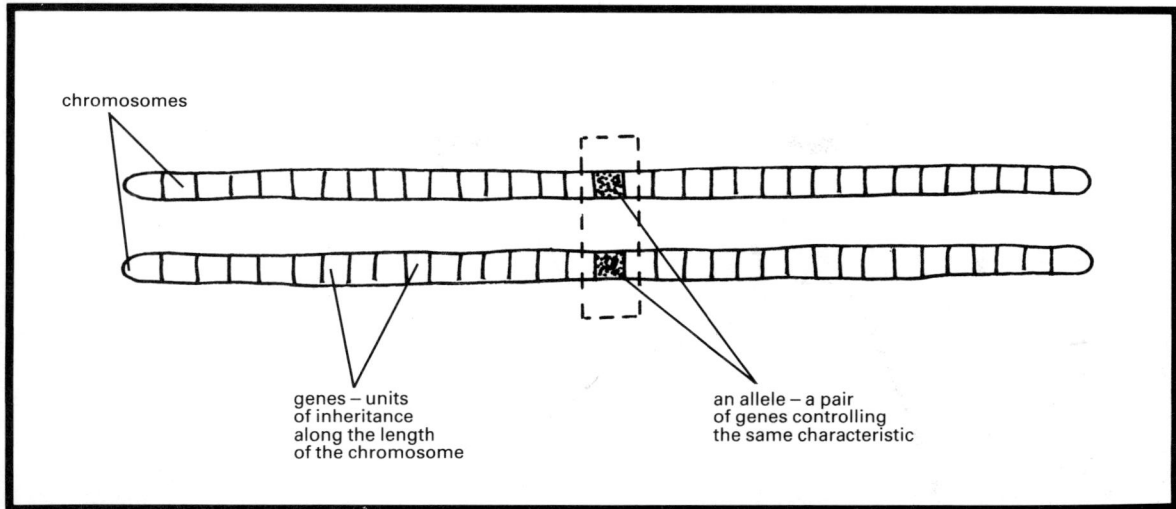

chromosomes

genes – units
of inheritance
along the length
of the chromosome

an allele – a pair
of genes controlling
the same characteristic

A pair of chromosomes (greatly simplified)

chromosomes side-by-side, genes of the same sort would lie alongside each other. Such a pair of genes, controlling one characteristic, is called an allele. Remember that the chromosomes are in pairs in normal cells and are single in gametes, so the same applies to genes: pairs in normal cells, singles in gametes.

Remember also, that all the members of one species have the same number and types of chromosomes, and consequently, the same number and types of genes. If we were to look at the same pair of chromosomes in two different dogs we would find that the genes controlling a particular characteristic were in the same position on each chromosome pair. It seems that the sequence of genes on the chromosomes forms a set of instructions which can be used to build up the individual animal or plant. So long as the instructions remain correct, then our dog's genes will always produce a dog, not a cat, a rat or a kangaroo!

Each gene is minute, so how can it carry its complicated instructions? It seems that the chemical composition of a gene is the key to its job as an information-carrier. Genes are made up of a chemical called DNA, which is variable in composition. Each small variation can be used like a sort of code, to tell the developing animal or plant what materials to manufacture to give it its unique structure. Variations in the chemical structure of the DNA can therefore make our dogs tall or short, long-haired or short-haired, black or white, and so on.

So, when two dogs mate, they pass on instructions in the form of genes which will decide the characteristics of their offspring. The genes are single in the gametes, so when the sperm and ovum join together and their chromosomes pair up again, one gene from each pair has come from the mother and one from the father. The result is a dog which looks like its mother and its father, but is also an individual, being a unique mixture of the two.

We cannot see genes by looking at the chromosomes. A molecule of DNA can be seen under a very powerful microscope as a long strand, but we cannot see the minute chemical variations which make up a gene. The most usual way of studying how genes actually work is to do special breeding experiments.

Experiments

A good way of studying genetics is to do breeding experiments. Plants, like these primroses, are often used because they are easier to deal with than many animals. Most plants can be grown quickly and in large numbers and they are easy to handle. The object of most genetics experiments is to breed together two individuals and find out what their offspring are like. In flowering plants this involves taking pollen from one flower and transferring it to the ovary of another. The seeds developing from this pollination have to be planted and grown to see what the offspring are like.

Problems arise at every stage of this experiment. First of all, you have to look for a characteristic which is easy to study, such as flower colour, seed colour or seed shape. You then have to make sure that the flowers have been properly pollinated. A clean, soft brush is used to brush pollen lightly from the anthers of one flower onto the ovary of another. The next problem is to prevent any other, unwanted pollen from getting to the flower. Before the pollinated flower opens, its own anthers are removed, to prevent it pollinating itself. After pollination, the flower is covered with a fine cloth bag which keeps out any other pollen grains, and insects which might carry them.

Gregor Mendel used these methods to study inheritance in garden peas. He used peas with easily identifiable characteristics, such as round and wrinkled seeds, tall and dwarf plants, green and yellow seeds. To make sure that his results were genuine, he pollinated huge numbers of plants and germinated hundreds of their seeds to find out what they were like. He repeated his experiments over and over again to make sure that any results he got were not just freaks.

All geneticists pay the same attention to detail when they do breeding experiments. An animal commonly used for these experiments is the fruit fly. Fruit flies can easily be kept in bottles with a sticky mixture of fruit, yeast and other foods in the bottom. They show certain easily recognizable inherited characteristics, such as eye colour, wing shape and body colour. The flies in the photograph, for instance, have different eye colours and they have no wings. They are easy to sex; males have a black-tipped body and females have stripes at the tip of the body. They are also easy to breed. The parents to be used in the experiment are anaesthetized with ether to make them easier to handle, and carefully transferred with a brush to a bottle containing food. When they recover from the effects of the ether they mate and the female lays dozens of eggs in the food. Two days later a tiny white grub hatches from each egg and feeds hungrily on the nutrient mixture. After three days the grub forms a pupa, and four days later a young fly emerges. The bottle becomes filled with young flies which are also anaesthetized and the different types are counted. For this experiment to be successful, the original female parent must be a virgin, that is, she must not have been mated with any other male. To ensure this, she is removed from her bottle as soon as she emerges, so she won't have had time to mate before the experiment begins.

One of the most interesting facts emerging from breeding experiments is the remarkable similarity seen in the way a whole range of animals and plants inherit characteristics. Gradually, geneticists have been able to piece together the evidence and work out some simple rules of genetics.

How to sex fruit flies

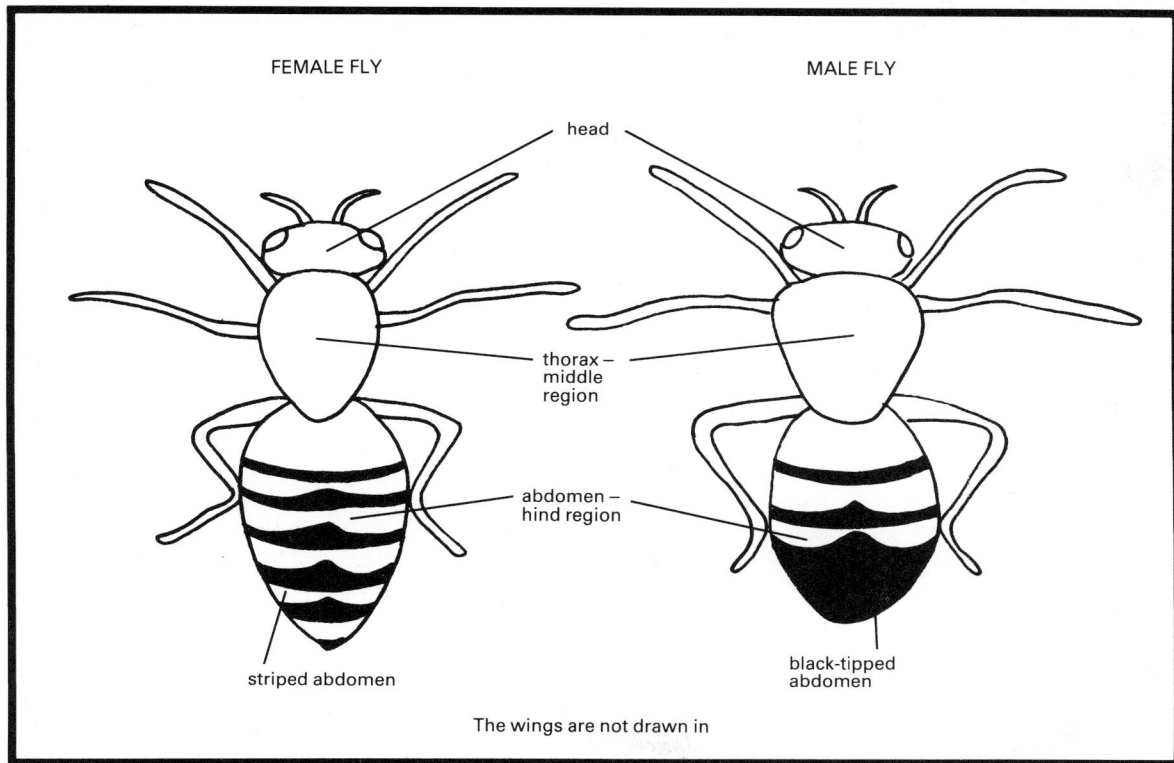

FEMALE FLY   MALE FLY

head

thorax – middle region

abdomen – hind region

striped abdomen

black-tipped abdomen

The wings are not drawn in

19

Dominance

Being able to roll your tongue is a trick you inherit from your parents. Some people can roll their tongue into a tube like this, others can't. Scientists studying inheritance within families have worked out a pattern which shows how tongue-rolling is passed from generation to generation.

We can't actually see the genes which control

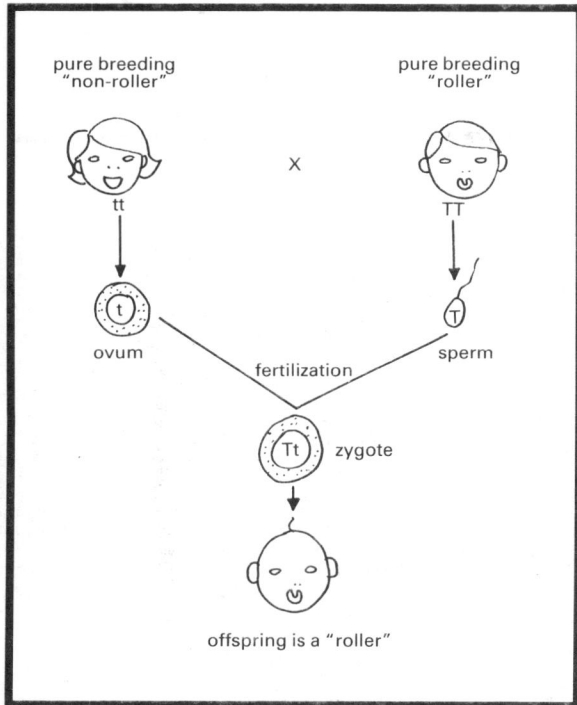

Inheritance of tongue rolling

tongue-rolling, so we have to try to imagine what happens and then test our ideas to see if they are right. We know that chromosomes exist in pairs in a normal cell, so the genes they contain will also be paired. Suppose that one pair of genes controls tongue-rolling ability. We can use a sort of shorthand code to describe these genes. Let's call a gene which enables you to roll your tongue T, and a gene which controls the inability to roll your tongue t.

Remember that each parent donates one gene of each pair to its offspring, which is therefore a mixture of the two. At fertilization, one parent may donate a T and the other a t, or both parents may donate a T, or both a t. So, the three possible combinations of genes in the pair are:

Tt, TT, tt.

The first one, Tt, is called a hybrid, and the other two, TT and tt, are said to be pure breeding. A person with genes TT is a tongue roller, a person with genes tt cannot roll their tongue, but what about Tt? In fact, the hybrid is also a tongue roller. So the T gene of the pair seems to suppress the effect of the t gene. Because of this, the T gene is called "dominant" and the t gene "recessive".

Many other gene pairs show this same dominance and recessiveness. For example, the "mousy" colour of wild mice, which is known as "agouti", is dominant over black colour. So, if the "agouti" gene is A, and the "black" gene is a, then aa produces black coat, and AA and Aa both produce agouti coat.

In garden peas, round seeds are dominant to wrinkled seeds so, if R is the gene for roundness and r is the gene for wrinkled seeds, then rr produces wrinkled seeds and both RR and Rr produce round seeds.

This pattern of dominance and recessiveness is not always so clear-cut. Some characteristics are controlled by more than one gene pair, so it is difficult to work out which is dominant and which recessive. The best way to study how characteristics are inherited is to pick a feature controlled by only one gene pair. In humans this is very difficult, since most characteristics are controlled by many gene pairs. However, tongue-rolling and the presence or absence of ear lobes are each controlled by one pair of genes only, and their inheritance can sometimes be traced right back through generations of a family.

The Basics

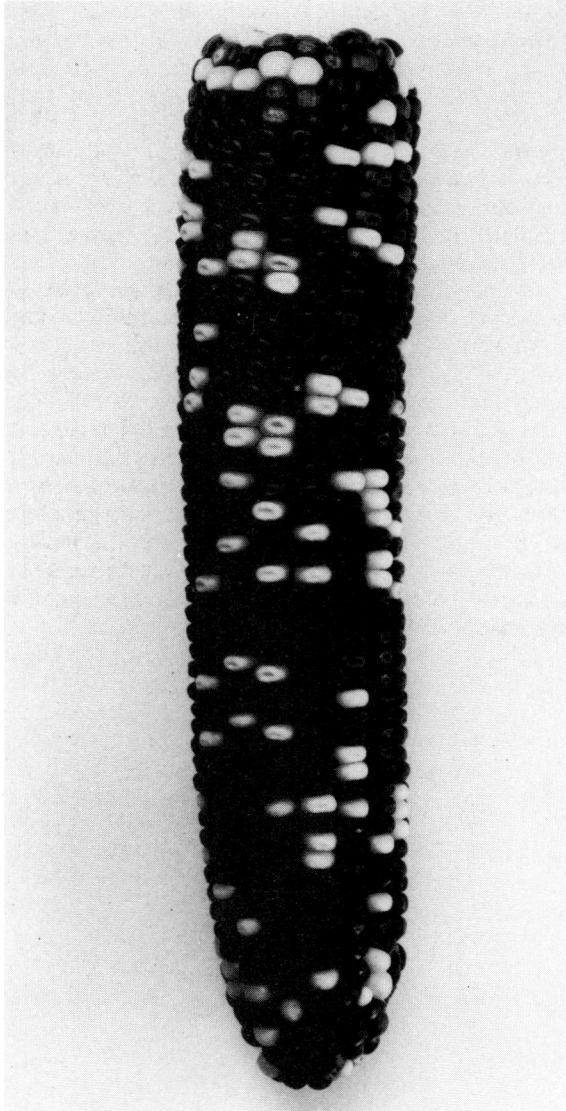

The surprising thing about inheritance is that it shows a similar pattern throughout the animal and plant kingdoms. The maize cob in the photograph is part of an experiment which illustrates this pattern of inheritance. If you count the different coloured seeds on the cob, you will find there are about three times as many dark ones as light ones. This is called a 3:1 ratio, and it appears again and again in genetics experiments.

Let's look at the experiment which produced these results. It began with two parents, a pure-breeding dark-seeded variety, and a pure-breeding light-seeded variety. These two were bred together, or crossed, and their offspring were studied. Surprisingly, all these offspring had dark seeds. These dark-seeded offspring were bred together (or self-fertilized) and the result was the maize cob in the photograph with the 3:1 ratio.

We can summarize this experiment in genetics "shorthand" as shown in the diagram. Here, we have called the gene controlling dark seeds D and the one controlling light seeds d. The parents were pure-breeding dark, or DD, and pure-breeding light, or dd. Remember that the genes are in pairs in the normal cells, but single in the gametes. So, all the gametes from the dark parent will contain the gene D, and all the gametes from the light parent will contain the gene d. When the gametes join at fertilization, the resulting offspring will therefore all be Dd. Now, all these offspring had dark seeds, so gene D must be dominant over gene d.

The next generation is produced by crossing together these dark-seeded offspring. Their gametes also contain only one gene, but half will contain gene D and the other half gene d. If you join together all these D and d gametes in

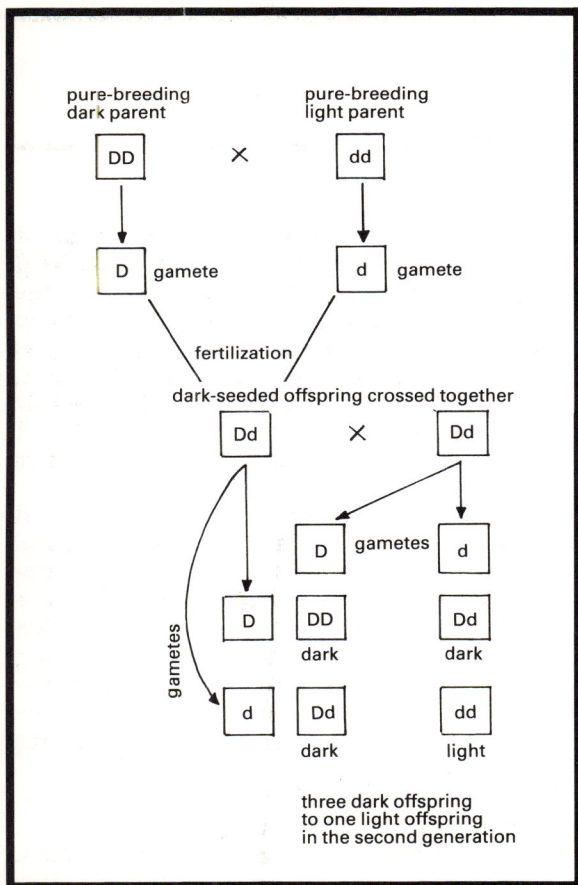

pure-breeding dark parent   pure-breeding light parent

DD   ×   dd

D gamete   d gamete

fertilization

dark-seeded offspring crossed together

Dd   ×   Dd

D gametes   d

D   DD   Dd
        dark   dark

d   Dd   dd
        dark   light

three dark offspring
to one light offspring
in the second generation

The 3:1 ratio in maize cobs

every possible combination, you get one out of four DD, two out of four Dd and one out of four dd. Since D is dominant over d, the DDs and the Dds will all be dark-seeded and the dds will be light-seeded, giving a ratio of 3 dark-seeded to 1 light-seeded.

The so-called "father of genetics", Gregor Mendel, knew nothing about genes, or even about the details of fertilization, but he recognized the 3:1 ratio and gave an accurate explanation of how it came about. He was an excellent scientist and mathematician, making sure the results of his experiments were valid by repeating them over and over again with different parents and different characteristics. He used huge numbers of plants for every experiment to rule out the possibility of his results being caused just by chance.

Mendel also made sure that his parent plants were pure-breeding by self-pollinating them over and over again to make sure they bred true. With all this attention to detail, Mendel's experiments became his lifetime's work. He wrote an article about his work in 1866, but it wasn't until after his death that its true value was recognized. In 1900, biologists were learning much more about sexual reproduction and inheritance, and dug out Mendel's work. They applied his results to the facts they had discovered and used them to form the basis of modern genetics.

Mixtures

Chickens aren't just egg and meat producers, they can be ornamental as well. The different breeds vary in size and shape as well as in the colour and length of their feathers. Another variable feature is the fleshy "comb" on the top of the head. These chickens show three of the four different comb types which are named after their shapes: pea comb, rose comb, single comb and walnut comb.

If you try breeding from these four different types, you find some very odd things happening. For instance, if you cross a pure-breeding pea-combed chicken with a true-breeding rose-combed chicken, you get all walnut-combed offspring. If you cross-breed these walnut-combed offspring amongst themselves you get some of every comb type appearing in their offspring. By careful experimenting, geneticists have discovered that there are two pairs of genes controlling comb shape. They are given the letters R and r, and P and p, and you can see

from the diagrams that different combinations of these genes give the different comb shapes.

It's not just chickens which make life complicated for geneticists. Some animals and plants have gene pairs which don't have a clear-cut dominant and recessive. This is called incomplete dominance. For example, if you cross pure-breeding red cattle with pure-breeding white cattle, you would expect all their offspring to be either red or white. Instead, the calves have a mixture of red and white hairs, which is called roan. If the genes are given the letters R and r, then RR gives red coat, rr gives white coat and Rr roan coat. If you cross the roan-coated offspring together you get calves in the proportions of 1 red coated : 2 roan : 1 white coated – not the expected 3:1 ratio.

Cats also give us an example of incomplete dominance. Black cats have genes BB controlling their coat colour and ginger cats have genes bb controlling theirs. But a hybrid cat, Bb, doesn't have a black coat, but the combination of black and ginger hairs we call tortoiseshell.

Another genetics puzzle also involves cats. A particular form of inherited deafness is more common in white cats with blue eyes. The inheritance of the three characteristics seems to be linked together, so this phenomenon is called linkage. It would seem that the genes controlling hearing, eye colour and coat colour are very close together on the same chromosome, so they are usually inherited together.

In fruit flies, curved wings and brown eyes are often inherited together; in other words, they show linkage. Fruit flies have only four pairs of chromosomes in each cell and the genes controlling brown eyes and curved wings must be close together on one of these pairs. By studying linked genes like these, geneticists have been able to discover the location of most of the genes on all the chromosomes. In other words, they have made chromosome maps. This is a long, complicated process, even with only four pairs of chromosomes. Think how difficult it would be to make maps of the 23 pairs of human chromosomes!

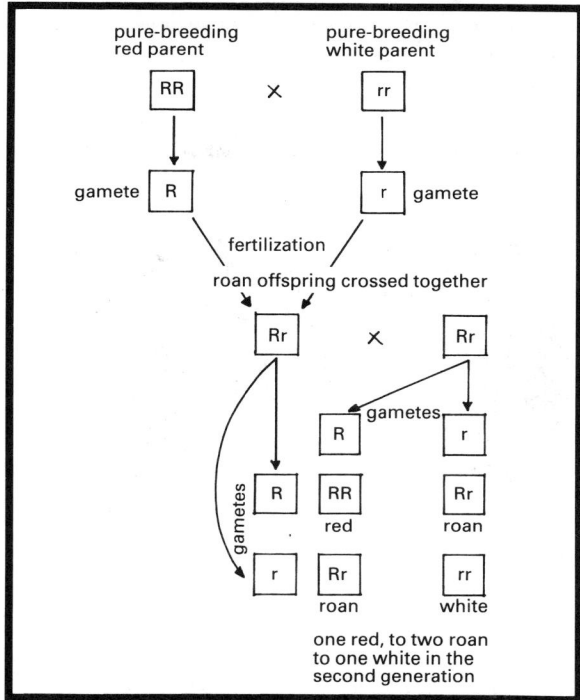

Inheritance in red and white cattle

The comb shapes of fowl and the genes controlling them

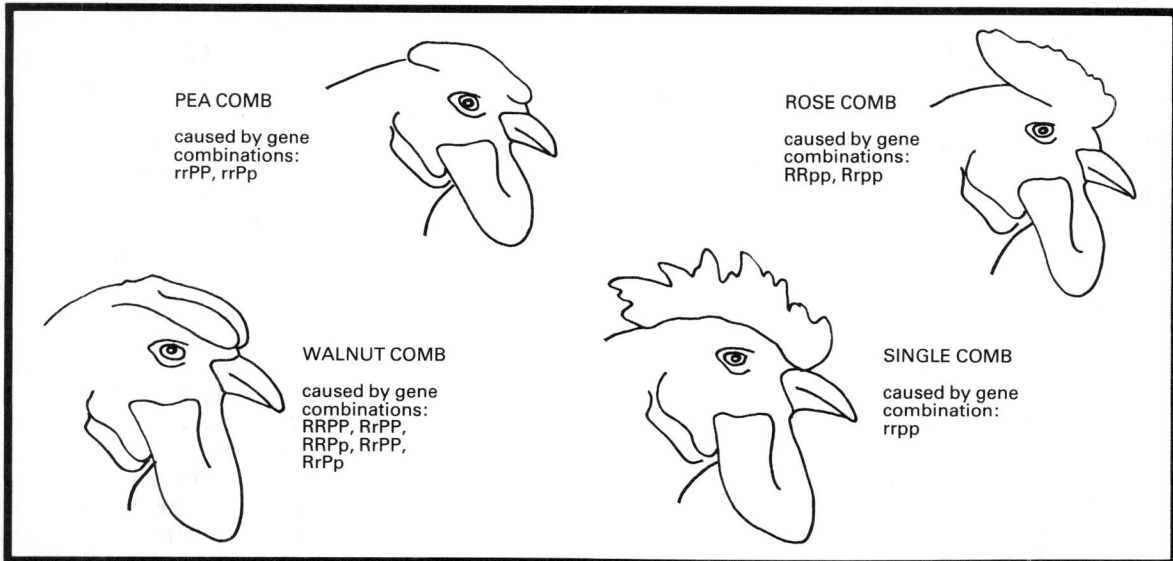

PEA COMB
caused by gene combinations:
rrPP, rrPp

ROSE COMB
caused by gene combinations:
RRpp, Rrpp

WALNUT COMB
caused by gene combinations:
RRPP, RrPP, RRPp, RrPP, RrPp

SINGLE COMB
caused by gene combination:
rrpp

He or She?

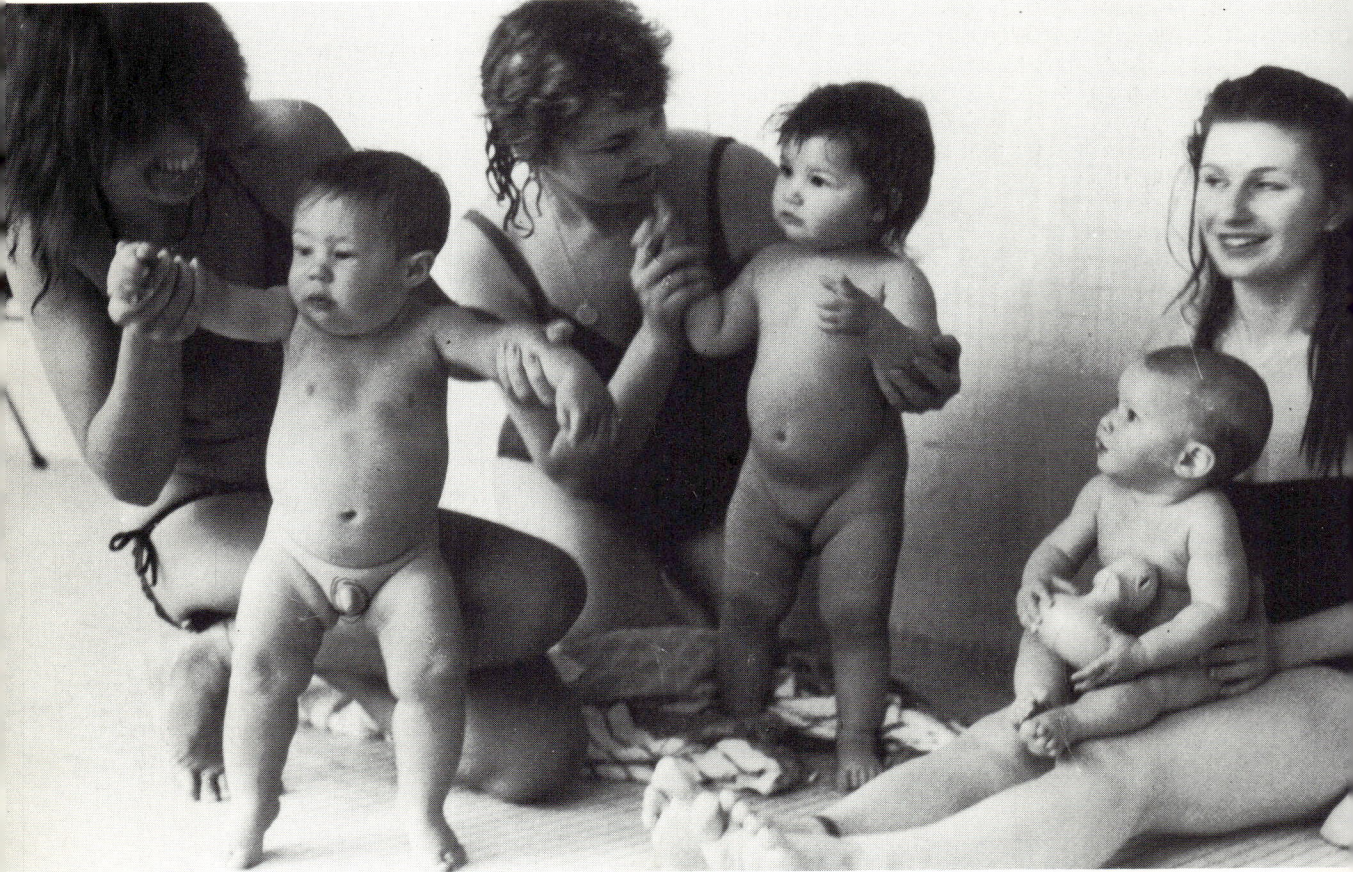

It's pretty obvious which of these toddlers is a girl and which a boy, but how did this difference begin? The sex of a baby is decided at fertilization, and it all depends on the chromosomes.

Remember that every human cell contains 23 pairs of chromosomes. Of these 23, one pair controls the person's sex. These sex chromosomes look different in males and females. In girls, the pair of sex chromosomes look like two similar Xs and so are called X chromosomes. In boys, one of the sex chromosomes is an X chromosome, but the other is much smaller and is Y-shaped, so it is called a Y chromosome. So, a female can be called XX and a male XY.

The X chromosomes contain genes which control many characteristics, not all linked to sex. The Y chromosome has only a few genes on it; it does not match the X chromosome.

When gametes are made, the chromosome pairs separate, one from each pair going into each gamete. This is true of all the chromosome pairs including sex chromosomes. When ova are made in a woman's ovaries, one X chromosome goes into each gamete. So, each ovum contains 22 ordinary and one X chromosome. When sperm are produced in a man's testes, the X and Y chromosomes separate into different cells, so half the sperm contain an X chromosome and half a Y chromosome.

At fertilization, if an X sperm joins with the ovum, the baby will be XX, or a girl. If a Y sperm joins with the ovum, the baby will be XY, or a boy. Since half the sperm have an X chromosome and half have a Y chromosome, there is a 50:50 chance of a baby being a boy.

When a baby is developing in the womb, it is surrounded by a sort of bag called the amnion, which is filled with a liquid, called amniotic fluid. The baby is constantly losing cells from its skin into this fluid. As we shall see, it is sometimes important that a mother-to-be knows whether she is expecting a girl or a boy. In this case, a syringe needle can be carefully pushed into her womb and a small amount of amniotic fluid sucked out. A study of the cells floating in the fluid will reveal whether there are the two X chromosomes of a girl, or the X and Y chromosomes of a boy.

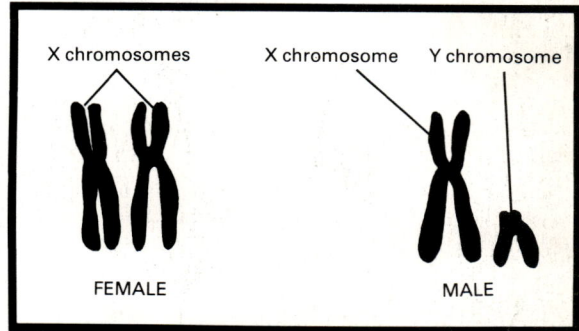

The sex chromosomes

How sex is inherited

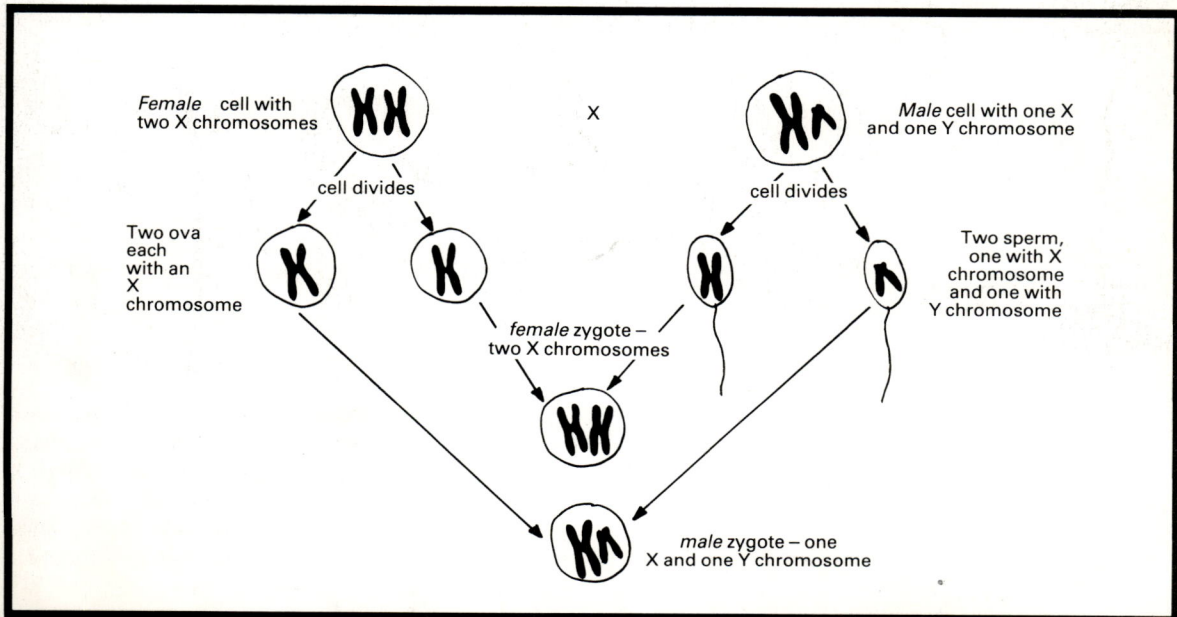

Sex Linkage

You are much more likely to be colour blind if you are a boy than if you are a girl. People who suffer from red/green colour blindness cannot tell the difference between red and green. To them, both colours look brown. This lady is having a colour blindness test. Each page of the book is covered with different-coloured spots. The dots form a pattern, often in the shape of a number, which can be distinguished by a normal-sighted person but not by a colour blind person. You can't see the pattern in this black and white photograph, because it depends on colour.

Red/green colour blindness is an inherited characteristic. The gene controlling it is on the X sex chromosome. It is a recessive gene, its dominant partner controlling normal colour vision. The Y sex chromosome does not have a corresponding gene controlling colour vision. Female cells have two X chromosomes, so they also have a pair of these "colour vision" genes. Male cells have one X chromosome and one Y chromosome, so they only have a single "colour vision" gene.

If we call the gene controlling normal colour vision N and the one controlling red/green colour blindness n, we can look at their effects in both men and women. Female cells could have the genes NN, Nn or nn. The first two, NN and Nn, would both produce normal vision, since N is dominant. Only nn would produce colour blindness. Male cells have only two possible gene combinations, since they only have one X chromosome. These are N, which would produce normal vision, and n, which would produce colour blindness. If we look at all possible combinations of these genes, it is much more likely for a boy to be red/green colour blind than a girl.

This sort of inheritance is called sex linkage; in other words, the inheritance of the characteristic is linked to the sex of the individual. Another sex-linked characteristic is the disease haemophilia. This is sometimes called "bleeders' disease" because the sufferer's blood is very slow to clot and even a minor injury can be very serious. It is controlled by a recessive gene on the X chromosome. It is inherited in the same way as red/green colour blindness, except that a girl with haemophilia has never been known. The gene controlling normal blood could be called B and that which controls haemophilia b. A woman with the genes BB obviously has normal blood. One with genes Bb also has normal blood, but is called a carrier, because she can pass the b gene on to

How colour blindness is controlled

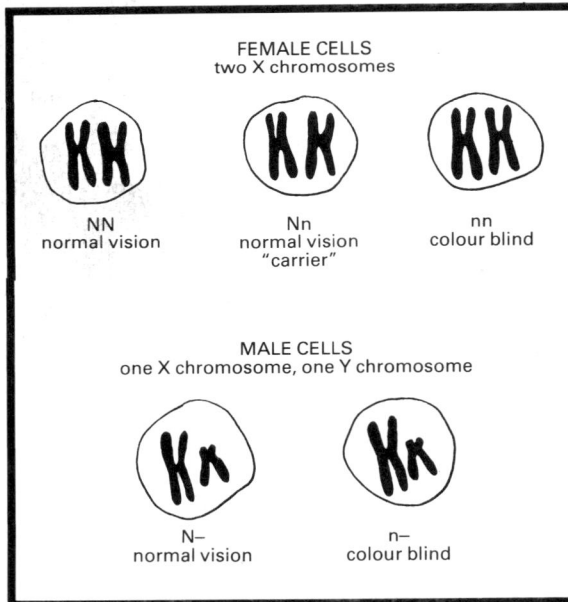

FEMALE CELLS
two X chromosomes

NN
normal vision

Nn
normal vision
"carrier"

nn
colour blind

MALE CELLS
one X chromosome, one Y chromosome

N–
normal vision

n–
colour blind

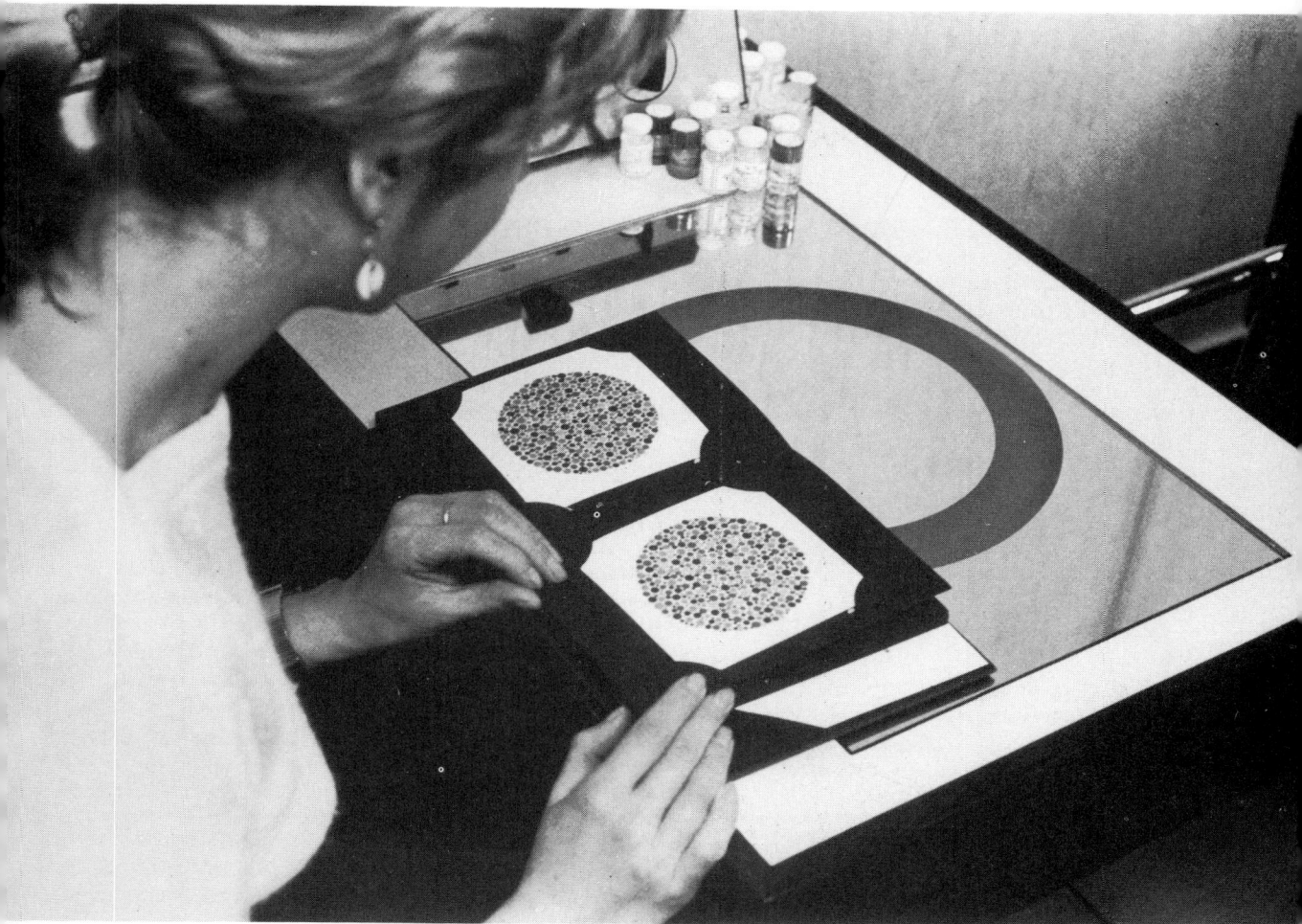

her children. A man with gene B on his X chromosome has normal blood, but one with gene b will have haemophilia. So, if a "carrier" woman marries a normal man, there is a 50:50 chance that any boys they have will suffer from haemophilia.

A number of men in Queen Victoria's family suffered from haemophilia. This family included the royal families of many European countries. Since it was the custom to marry within this extended royal family, the disease was more likely to crop up. Only by marrying outside the European royal families was it possible to reduce the chance of inheriting the disease.

But sex linkage doesn't only happen in humans, it occurs in other animals as well. For instance, tortoiseshell cats are much more likely to be females than males. The tortoiseshell colour is produced by a dominant and recessive gene on the X sex chromosomes, so it is only possible in females, who have two X chromosomes, not in males, who have only one.

Twins

Out of every 80 births in Britain, one is likely to be twins. The young men in the photograph are identical twins. They are very difficult to tell apart and their behaviour and personalities are likely to be similar as well. Some twins are only as similar as ordinary brothers or sisters and are called non-identical, or fraternal twins.

But how do twins come about? To find out we have to go back to just after the mother's ovum was fertilized. After fertilization, an ovum normally divides into two, then four, then eight, then 16, and so on. But sometimes, the first

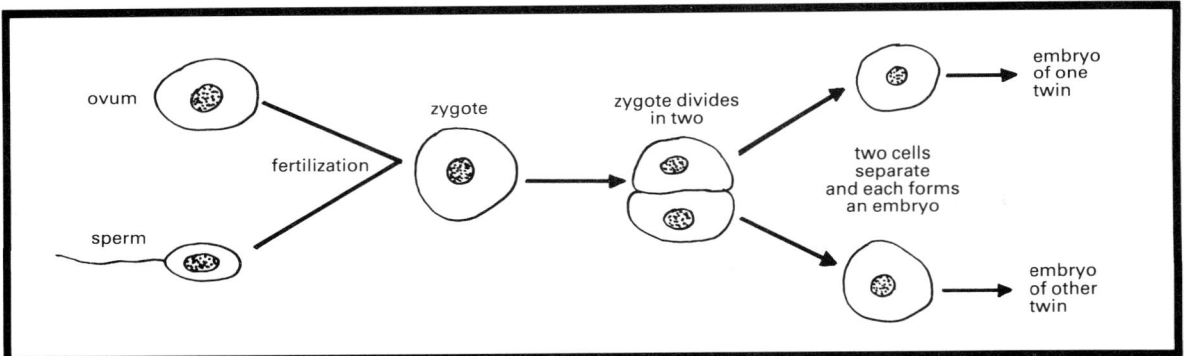

How identical twins are produced

How fraternal (non identical) twins are produced

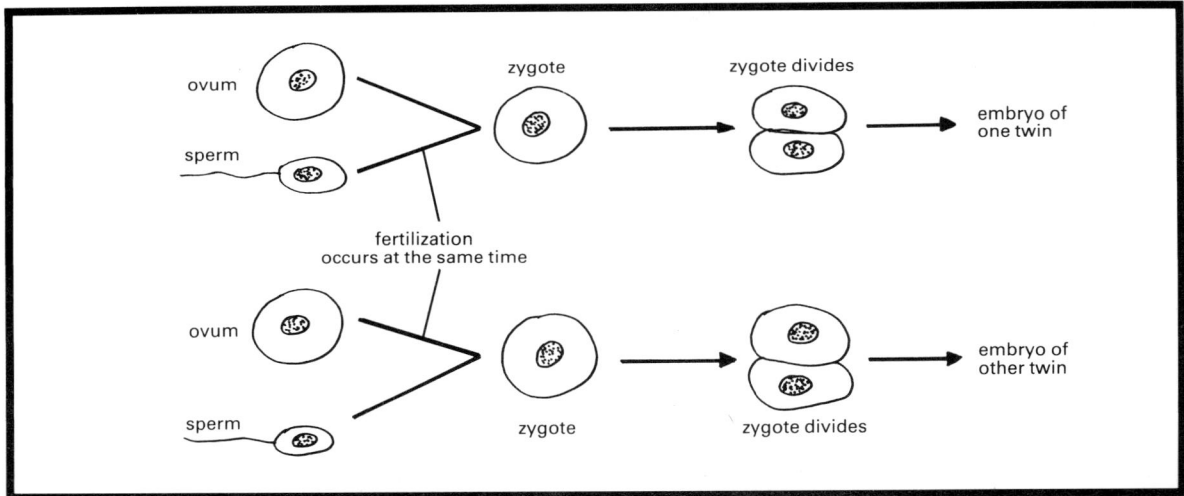

two cells produced don't stay together; they separate and each develops into a baby. These become identical twins and because they come from the same cell they are always of the same sex.

Identical twins look the same because their cells contain the same genes. Remember that the fertilized ovum, or zygote, contains half its chromosomes from its mother and half from its father. When it divides in two these chromosomes are exactly duplicated in each cell. So, each of the two cells contains exactly the same chromosomes, and therefore, the same genes as the other. With the same combination of genes, the cells are going to develop into people with the same characteristics.

However, it is possible, although rather difficult, to tell identical twins apart. Their genes may be exactly the same, but slight differences can arise because of differences in their surroundings. For instance, one twin is often a bit larger and stronger than the other. This may be because he had a better supply of oxygen from his mother's blood in the womb than his twin. The conditions which produce these differences are called environmental factors and they are nothing to do with the twin's genes. If the twins grew up in *exactly* the same conditions, then they would look exactly alike.

Fraternal twins do not look exactly alike, and they may be of the same or different sexes. They are the result of two separate ova being fertilized at the same time. Usually a woman produces only one ovum at a time, from alternate ovaries. But occasionally, both ovaries make an ovum at the same time. If both these ova are fertilized by different sperm, then they will develop into fraternal twins.

Fraternal twins come from two different ova and two different sperm. The genes in these ova and sperm will be very slightly different although they come from the same mother and father. So, fraternal twins will inherit some characteristics from their mother and some from their father, but not necessarily in the same combination. They will look no more alike than normal brothers and sisters. In fact, the only thing about them which will be exactly the same is their age.

Human Genetics

Human beings show infinite variation. We have different skin and eye colours, varying hair colours and textures and different-shaped features. We each inherit some characteristics from both parents because we possess a mixture of genes from each of them. No other person, not even our brothers and sisters, contains the same mixture of genes, so no one looks exactly like anyone else. Many of our characteristics are controlled by more than one pair of genes, and their genetics are very complicated. Some features, however, are controlled by only one gene pair.

One such feature is the presence or absence of a fleshy outgrowth, or ear lobe, at the base of the ear. The absence of ear lobes is controlled by a recessive gene, so far more people have ear lobes than not. Similarly, a recessive gene controls the inability to taste a chemical called PTC. Suppose the gene controlling the ability to taste PTC is called P and that controlling the inability to taste it is called p. Then PTC "tasters" have the genes Pp or PP and non tasters have the genes pp.

Other inherited characteristics are not so simple to study. The two examples already mentioned have only two alternatives; either you have ear lobes or you don't, either you can taste PTC or you can't. But for some inherited characteristics we fall into more than two groups. Take blood groups, for instance. There are four blood groups in the so-called ABO system; group A, group B, group AB and group 0. These groups are inherited in such a way that it is quite likely for our blood group to be totally different from either of our parents'. For example, if two people with blood groups A and B marry, they are quite likely to have children with blood group O. However, if either of the parents has blood group AB, then it isn't possible for their children to have blood group 0. This fact has sometimes been used in legal cases, perhaps where babies have been mixed up in hospital, or when a man claims he is not the father of a particular child.

Human genetics is difficult to study because you can't do breeding experiments on people. You can't tell people whom to marry, and you can't order them to have children. On the other hand, you can look at family trees and trace how different characteristics have been inherited. The trouble is, most ordinary people know nothing about their distant ancestors, so we have to look at well-documented family trees, such as those of royal or aristocratic families. In this way, we have been able to unravel the genetics of Queen Victoria's family who were afflicted with haemophilia.

However, geneticists can look at how inherited characteristics occur in large populations of unrelated people and, by statistically analysing their results, can discover more about human genetics. This is called population genetics. And because they are genetically exactly the same, identical twins can be studied to try to unravel the mysteries of human genetics. By measuring how much identical twins actually differ, we can find out how much of their make-up is controlled by genes and how much by other factors.

Genetic Diseases

This boy has muscular dystrophy. His muscles are gradually being replaced by fatty tissue, so they are unable to support his body or to move his limbs properly. His legs aren't strong enough to bear his weight and he will become weaker and weaker until he is unable to do anything for himself. Muscular dystrophy is an inherited disease, passed on from mother to child. It is one of many inherited diseases, some of which are fatal or severely crippling.

Sickle cell anaemia is an inherited disease suffered mainly by negroes. The red cells in the blood are responsible for carrying vital oxygen to all parts of the body. Normally they are flexible and disc-shaped, but in sickle cell anaemia they are fragile and crescent-shaped. These sickle cells tend to break up easily, and block the tiny blood vessels in the body, especially in the internal organs. The sufferer is weak and short of breath. Also his liver and spleen are likely to become diseased, because these organs are responsible for dealing with broken down blood cells.

Cystic fibrosis is the commonest serious inherited disease of white children. It is caused by a build-up of slimy mucus in the lungs, the mouth and gut. The child fails to gain weight, in spite of a healthy appetite, and has frequent attacks of bronchitis. The disease can be treated by physiotherapy for half an hour twice a day. This loosens the mucus which causes the symptoms, but, of course, it would be better if the mucus production could be stopped altogether.

Some genetic diseases are caused by abnormal chromosomes. Some people have an extra chromosome in their cells and suffer from Down's syndrome or mongolism. They are mentally deficient and have a characteristic round head, almond-shaped eyes and thick, straight hair. Down's syndrome seems to be caused by a "mistake" happening at the time the ovum is made in the ovary. It is much more common in the babies of older women than of young ones, perhaps because the older ovaries are getting a bit "worn out".

Nowadays tests can be made which can predict whether a child will suffer from an inherited disease. In the case of sickle cell anaemia, the parents' blood can be tested to see if they are both carriers of the disease. They may then decide not to have children, since there is a one-in-four chance that they would have the disease. Other inherited diseases can only be detected during pregnancy. Some diseases, such as spina bifida, can show up in a chemical test on the pregnant mother's blood, but for others, the material for testing must come directly from the baby. Amniocentesis, studying the cells shed by the baby into the fluid surrounding him, can show up chromosome abnormalities, such as Down's syndrome. It can also detect the sex of the baby, a fact which is important if a sex-linked disease such as haemophilia runs in the family. An even more complicated test involves very carefully removing some of the blood from the baby's umbilical cord. Although there is some risk to the baby, this method reveals genetic diseases very early on in pregnancy. If the baby is affected, the parents must decide whether the mother should have an abortion, an operation to remove the baby from the womb. The earlier this decision is made, the better.

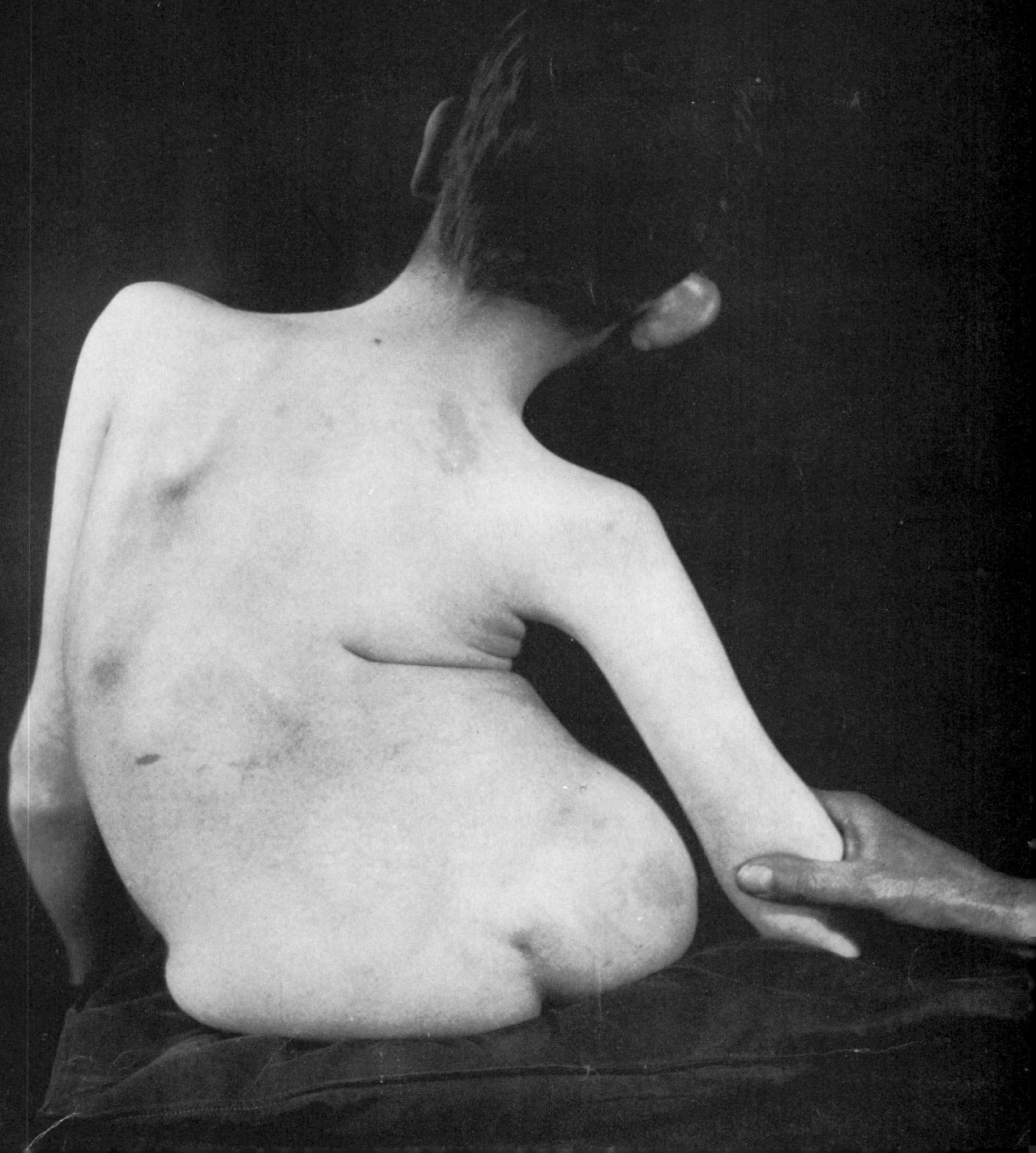

Animal Breeding

These prize cattle are the result of a partnership with man stretching back to Neolithic times. Ever since man stopped hunting his food and began farming, he has used animals to provide him with food, hides, wool, transport and companionship. But Neolithic domestic animals were a far cry from our modern varieties, which have been carefully bred to meet our needs.

Over the thousands of years since they were first domesticated, these animals have gradually changed from the wild type, suited to life in natural surroundings, to the specialized breeds we know today. Cattle may be good milk producers, such as Jerseys and Guernseys, or produce good beef, such as Herefords and Shorthorns. The black and white Friesian cattle are doubly valuable, since they are good for both meat and milk production. Another versatile animal is the dog, which has evolved from its original, wolf-like ancestor into the hundreds of breeds we know today. No truly wild dogs exist now, but the originals were probably very much like wolves. From this basic wolf-like stock have arisen all the different breeds of dogs, each suited to a specific purpose. Some dogs, such as bloodhounds, can track prey by scent; others, such as retrievers, can bring back shot game. Welsh collies round up sheep, huskies pull sleds and St Bernards rescue mountaineers in difficulty. And, of course, some dogs, such as King Charles spaniels, have been bred purely for ornament.

How has this change from wild to domestic breeds taken place? Neolithic man probably began the process unconsciously. He would have picked out the animals most useful to him, such as the cattle producing the most milk, or the pigs providing the best meat, and used them for breeding. Continually picking out breeding animals with the best characteristics would cause a gradual change to animals better suited to man's needs. This process is called artificial selection.

We still use artificial selection today, but in a far less haphazard way than early man. Not only do we pick out the most desirable animals to breed from, we also look back at their pedigrees, to see how their ancestors have performed. For instance, we can study a bull's pedigree, and see if his female ancestors were good milk producers. We can then be almost certain that his daughters will be good dairy cattle as well.

We can also decide which animals will breed together. In some cases, it is not a good idea to breed from close relatives. Such inbreeding sometimes produces faults, such as hip weaknesses in many breeds of dog. So, animal breeders consult the pedigrees of the animals they hope to mate, to make sure they aren't closely related. Sometimes, especially with cattle, the two animals being mated don't even have to meet. Instead, sperm is collected from the male and placed artificially inside the female, a process known as artificial insemination, or A.I.

The trouble with many domestic breeds is that they may possess undesirable features along with the useful ones. Nowadays, animal breeders are showing an interest in old breeds which may have desirable characteristics such as vigour and resistance to diseases. These old breeds can be used to breed with modern varieties and improve them. In other words, they can provide a "gene pool" which can be drawn on to introduce useful features into a modern breed.

Plant Breeding

Thanks to the plant breeder, we have a wonderful variety of plants to choose from when we're planning a garden. These roses are just a few of the hundreds of varieties with a whole

range of colours, scents, shapes and sizes. Each year, more new roses are introduced, as the breeders try to make varieties just that bit different from the others. But every new variety is the end product of years of careful work.

The plant breeder perhaps has an easier task than the animal breeder. Like the animal breeder he can use artificial selection, picking out the plants which show the characteristics he's looking for, and breeding from them.

However, unlike the animal breeder, he doesn't usually have to look for a male and a female to breed from. Most flowering plants, including roses, contain both male and female organs, so the plant breeder can self-fertilize them. He does this by brushing pollen from the flower onto the sticky stigmas of its own ovaries. He prevents pollen reaching it from other flowers so that he knows any seeds he collects are the result of self-fertilization (see page 18). From each generation of plants he picks out the ones which show the characteristics he wants, and gradually he develops a pure-breeding new variety.

But the plant breeder has other tricks up his sleeve. New varieties of plants can also be produced by crossing together two different species. For instance, wheat plants can be crossed with other types of grass to give a better harvest, or resistance to diseases, or good growth in poor soil. In fact, this is one way in which modern bread wheat has changed from wild wheat. The wild wheat plant gives a poor yield and its seeds tend to drop too easily when it is harvested. When early man first grew wheat, he took his crops with him when he moved. The wheat plants naturally crossed with local grasses and the resulting plants, called hybrids, produced more, firmly attached seeds than the wild variety. Nowadays, plant breeders can choose which plants they want to cross to give a good hybrid, a process known as hybridization.

Sometimes, plants will appear in the wild which have larger flowers, or bigger fruits or seeds than others of their species. Examination of their cells shows that these larger plants have twice, or maybe three times the normal number of chromosomes. These plants can be bred from to produce larger flowers, or better-yielding crops. These freaks of nature, or polyploids, are responsible for larger apples, potatoes, wheat and sugar beet, to name just a few. Plant breeders can also produce polyploids artificially, often using a chemical called colchhicine, to develop better-yielding crop plants and larger, showier flowers.

Mutations

This white rabbit had normal parents. Its white colour is due to a mutation: a sudden change in the genes, which can be passed on to its offspring. Animal and plant breeders have often used such mutants, or "sports", to breed new varieties.

Mutations usually happen when the gametes are being produced in the sex organs. For instance, there may have been a fault in the production of the ova in the ovaries of the white rabbit's mother. Alternatively, the mistake may have been in the father's testes, which produced a faulty sperm. Either way, the white rabbit has a white, or albino coat, which was not present in its parents, but which it can pass on to its offspring. Since it is the chromosomes which carry the information required to produce the white rabbit, it is here we must look for the cause of the mutation.

If you imagine each chromosome as being a long string of genes, you can visualize the sorts of mutations that might occur. Part of a chromosome may be missing, or a little bit be added to it. Sometimes a piece of the chromosome is muddled up, and sometimes it seems to be completely scrambled. Usually, though, mutations are caused by only tiny changes in the chemical make-up of the chromosome. This child suffers from Down's syndrome, or mongolism, which is caused by an extra complete chromosome in her cells.

As we have seen, Queen Victoria's family suffered from haemophilia, a disease affecting males in which the blood clots only very slowly. It is likely that this disease was caused by a mutation in one of Queen Victoria's parents' sex cells, since none of her ancestors had haemophilia. Once the mutation had appeared in the royal family's genes, it was passed on to later generations. Only by marrying outside the immediate family was it possible to eliminate the disease from the family.

All sorts of other mutations have been studied in a variety of animals and plants which have been used for breeding experiments. In fruit flies, different coloured eyes, such as white, brown, apricot and purple, are caused by mutations. Corn seedlings with fine white stripes on their leaves are the result of a mutation, so are pigeons with coloured flecks on their feathers. There is even a mutation which makes mice move in such an odd way that they are called 'waltzing' mice. Often a mutation has such a disastrous effect that the organsm doesn't live. This sort of killing mutation is said to be 'lethal'.

Nobody really knows what makes mutations happen in nature, but scientists can induce them in the laboratory. Anything which causes mutations is called a mutagen. Some chemicals are known to be mutagens, and so are high doses of X-rays. In the early 1970s, a food preservative called AF-2 had to be withdrawn from use because it was found to be a mutagen. But the mutagen with the most horrifying effects is radioactivity. Exposure to radioactivity in extremely high doses, as after the atom bomb attacks on Hiroshima and Nagasaki in 1945, causes death or cancer, but in lower doses, it can increase the chances of mutations when sperm and ova are made in the testes and ovaries. These faulty sperm and ova can then pass on their mutations to future generations, producing deformed babies for many years after the exposure to radioactivity.

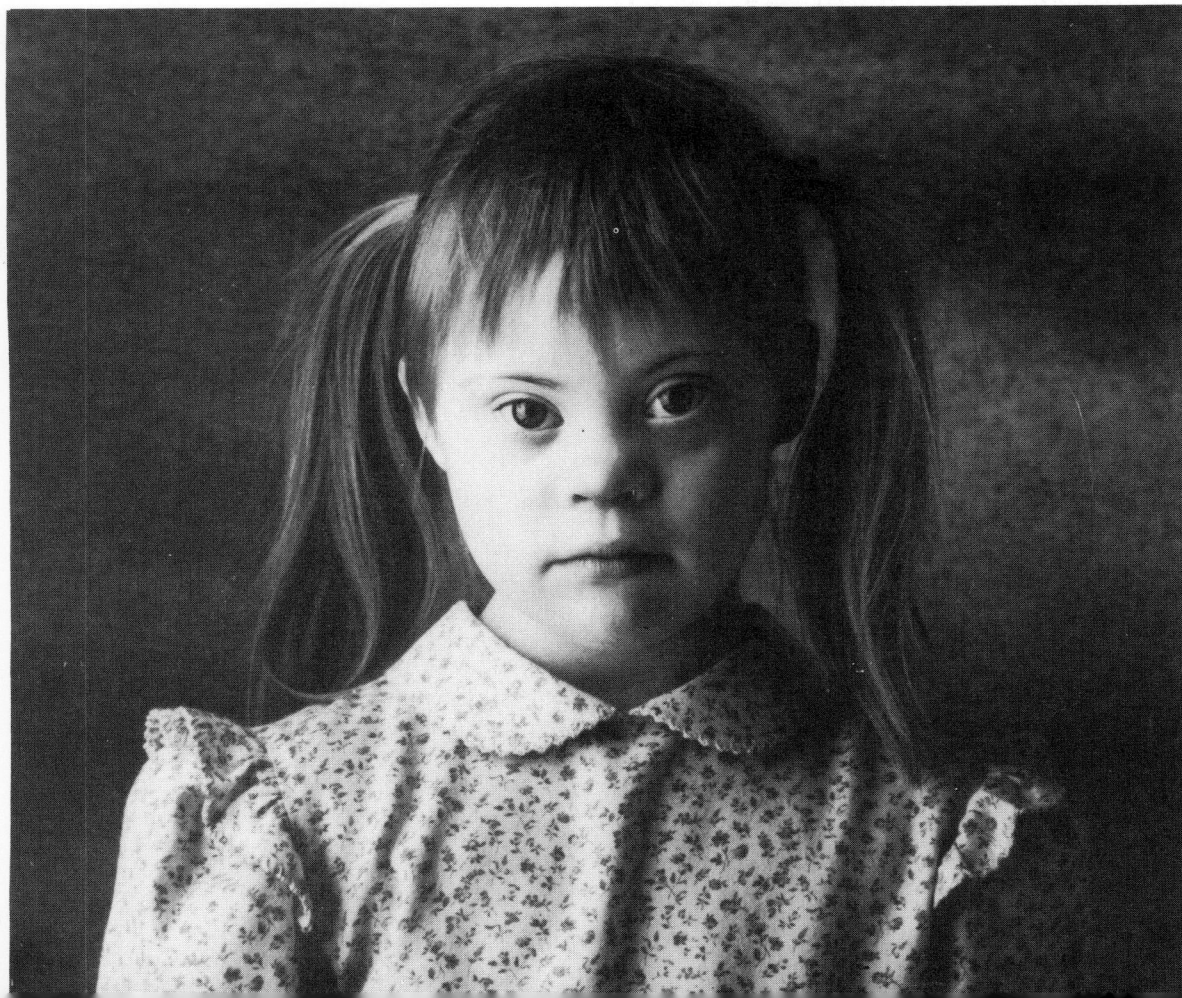

Evolution

Things have changed since this fossil dinosaur, *Triceratops*, roamed the earth. Scientists believe that living things are constantly, if slowly, becoming more complex. This change from simple living things to complex animals and plants is called evolution.

By looking at fossils, we can see how animals and plants have evolved. The deeper you dig, the older the fossils get, so we can build up a picture of evolution, called the fossil record.

The fossil record

ANIMAL LIFE	MILLIONS OF YEARS AGO	PLANT LIFE
Modern man		Modern types
	0.01	
Woolly animals equipped to withstand Ice Age		Modern types
	1.8	
Man's ancestors –"ape-men"		Modern types
	6	
Apes and grazing mammals		Grassland
	22.5	
Early apes. Many modern mammals appeared		Many flowering plants
	38	
Early elephants and horses		Most modern types appeared
	55	
Mammals evolving rapidly		Flowering plants taking hold
	65	
Dinosaurs dying out		First flowering plants
	141	
Dinosaurs dominant. First birds		Conifers
	195	
First dinosaurs. First mammals		Luxuriant forests of conifers
	230	
Reptiles increasing. Amphibians decreasing		First conifers
	280	
Many amphibians. First reptiles		Swamps of ferns and mosses. Later formed coal
	345	
Abundant fish. First amphibians		Land plants spreading
	395	
Large armoured fish and sea scorpions		First plants emerged on marshy land
	435	
First fish. Abundant marine invertebrates		Seaweeds and other algae
	500	
Large variety of marine invertebrates		Seaweeds and other algae
	570	

Some organisms don't fossilize well, but other sorts of evidence have filled in the gaps in the fossil record. We believe that animals began as single cells living in the water which covered much of the Earth 3000 million years ago. Then came the invertebrates, or animals without backbones, like jelly fish, worms and insects. These were followed by the animals with backbones, or vertebrates, starting with fish, then amphibians, then reptiles, like the *Triceratops*. The Age of the Reptiles began 195 million years ago, when dinosaurs lived in all sorts of different surroundings; there were flying dinosaurs, swimming dinosaurs, fierce carnivorous dinosaurs and meek, vegetarian dinosaurs. But 130 million years later, these versatile reptiles had all but disappeared, and in their place were the birds and the mammals.

Alongside the changing animal kingdom, plants were evolving as well. They, also, started as single cells which gave way to simple, many-celled plants, such as seaweeds. Then came mosses, followed by ferns, some of which grew into large, tree-like plants. These evolved into the cone-bearing trees, or conifers, until finally, 141 million years ago, the flowering plants made their appearance.

We can't go back in time to see exactly how evolution took place, but we can use our knowledge of genetics to make a few guesses. Suppose a mutation made an animal or plant more able to survive than the rest of its kind. This mutant would be more likely to produce offspring than the others, so passing its advantage on to the next generation. Gradually, the original plants or animals would die out, to be replaced by the mutant ones.

Let's look at how this might work in real life by looking at our mutant white rabbit. In normal surroundings the dull brown coat of the

rabbit camouflages it in hedgerows against attack by enemies. A white rabbit would be at a distinct disadvantage here, because its enemies would see it easily. But what if the climate changed and the rabbits had to escape their enemies in the snow? Now the white rabbit would be better camouflaged and better able to survive. The longer it survived, the more offspring it would produce and the white ones would also survive better and produce more offspring. Gradually, the whole rabbit population would change to the white-coated variety. This change, by means of "the survival of the fittest", is called natural selection.

Natural selection relies on the fact that mutations can be passed on from generation to generation. Most mutations are not advantageous, so they tend to disappear, but any which make an animal or plant better adapted to their conditions of life, could lead to evolution. And evolution hasn't stopped, it is continuing all the time. In another hundred million years or so, man may be extinct, having been replaced by more complex, better adapted, more intelligent life forms.

The Future

This strange animal is neither a sheep, nor a goat. It is a cross between a sheep and a goat, with some of the features of both parents. Animal and plant breeders use experiments like this to introduce useful characteristics into a breed. For instance, the goat side of this animal may have genes which make it resistant to diseases. By breeding the sheep/goat hybrid with other sheep, the useful gene may be incorporated into future generations of sheep. This is one form of "genetic engineering"; or changing an organism's genetic make-up for our own good.

Another way of changing genes is by artificially producing mutations. We could expose an animal or plant to X-rays and pick out

mutants with desirable characteristics to breed from. This has been tried successfully with a mould fungus which produces the antibiotic penicillin. A mutant was discovered after X-ray treatment, which produced a far greater quantity of penicillin than the original mould. Unfortunately, most mutations are harmful to the organism, and it would only be a matter of chance that a useful mutant was produced.

What geneticists would really like to do would be to pin-point harmful genes and change them for the better. In some experiments, viruses have been used to introduce genes into cells. Viruses are tiny micro-organisms which cause diseases by "latching on" to their victim's chromosomes. It is possible to attach a useful gene to a virus and to get it to infect a cell. The useful gene then becomes incorporated into the cell's chromosomes. It may soon be possible to "block" the effect of a harmful gene by using chemicals, or even to knock it out completely using a laser beam. The trouble with all these methods is that other genes may also be affected, with possibly disastrous results.

This is a "test tube baby". Nine months ago, doctors removed an ovum from his mother's body and added his father's sperm to it, so it was fertilized in a test tube (or more likely in a shallow dish). Once fertilized, it was carefully placed in the mother's womb, where it grew in the normal way. It all seems very complicated, but it's obviously been worth it for the proud parents who would otherwise have been unable to have children.

The trouble with this procedure is that an irresponsible geneticist could begin deciding who a child's mother and father should be. It has been suggested that the sperm and ova of famous people should be used to produce superior beings. Of course, this may not be successful, because the sperm and ova may not contain exactly the right combination of genes.

However, some scientists seem to have overcome this problem. They have found that it is possible to persuade any cell in the body to reproduce itself and behave exactly like a fertilized ovum. In the future, it may be possible to take any body cell and implant it in a woman's womb, where it would grow into a normal baby. Since every body cell contains exactly the same chromosomes and genes, the baby would be exactly like the person who donated the cell. This process is called cloning and it could have horrifying applications. You may think it would be marvellous to be able to produce more of your favourite pop star, or sporting personality, but imagine what would happen if someone used cloning to make hundreds of criminals. Obviously, geneticists have a tremendous responsibility to use their knowledge to benefit mankind or not to use it if it could be harmful in any way.

Glossary

anaesthetize: to use a drug which makes a person or an animal unconscious.

antibiotic: a chemical, made by a micro-organism, which can be used to kill other, disease-causing micro-organisms.

carnivorous: meat, or flesh-eating.

cell: the basic unit of living things. All animals and plants are made up of minute "building blocks" called cells.

cell wall: a strong, fairly rigid outer layer only found in plant cells.

cytoplasm: the part of a cell, outside the nucleus, in which all the chemical reactions take place to keep the cell alive.

embryo: an animal or plant which is growing and developing but is not yet fully formed.

fossil: the remains of an animal or plant which lived in the past. Many fossils are formed by the hard parts of the animal or plant turning to stone.

gamete: a single cell, produced in the sex organs of an animal or plant, which joins with another gamete to begin the life of a new animal or plant.

mammal: an animal with a backbone, whose body is covered with hair or fur, and which suckles its young with milk.

micro-organism: an animal or plant which is so small it can only be seen using a microscope.

mineral salts: chemicals in food which are needed in small quantities to keep the body healthy.

mucus: a sticky material which helps keep the inner surfaces of the body moist.

nestling: a young bird which has not yet left the nest.

nucleus (plural:nuclei): the part of the cell which is responsible for controlling the living processes of the cell, and for reproduction and inheritance.

organism: a living thing, either animal or plant.

oxygen: a gas, present in the air, which is needed by all living cells to keep them alive.

resilient: resistant to harsh conditions.

reproduction: a process of living things whereby new individuals are produced, ensuring the survival of the species.

reptile: an animal with a backbone, whose skin is covered with dry, leathery scales.

species: a group of animals or plants which have similar characteristics and which can inter-breed.

womb: the part of a female mammal's body in which a baby mammal grows and develops before it is born.

zygote: the cell which is produced when two gametes join together and which divides over and over again to produce a new animal or plant.

Book List

Auerbach, Charlotte, *Heredity*, Oliver and Boyd 1965
Cohen, N., *Discovering Genetics*, Longmans 1982
Jennings, T., *The Young Scientist Investigates Flowers*, Oxford University Press 1981
Mackean, D.G., *Introduction to Genetics*, 3rd ed., J. Murray 1977
Ward, Brian R., *Birth and Growth*, Franklin Watts 1983

Index